International R

The Horse Rainbow
An Equine Genetic Guide
1st Edition

By Melanie Patton

This book is from the Hippology: Equine Science Textbooks regulated by the International Riding Academy
Course Textbook for EQUINE GENETICS

The Horse Rainbow: An Equine Genetic Guide

Copyright © 2011 by Melanie Patton. All rights reserved.

Published and produced by the **INTERNATIONAL RIDING ACADEMY**

No part of this book may be reproduced, stored in a retrieval system or transmitted in any form or by any means, electronically, mechanical, photocopying, recording, scanning or otherwise.

Limit of Liability/Disclaimer of Warranty: while the publisher and author have used their best efforts in the preparation of this book, there is no representations or warranties with respect to the accuracy or completeness of the contents of this book and specifically disclaim any implied warranties of merchantability or fitness for a particular purpose. The advice and training strategies may not be suitable for a specific reader's situation and he should consult with a professional where appropriate. Neither the publisher nor the author are reliable for any loss of profit or commercial damages, including but not limited to special, incidental, consequential or any other damages.

Manufactured in the United States of America
10 9 8 7 6 5 4 3 2 1
1st Edition

Illustrations & diagrams by Melanie Patton

The Horse Rainbow: An Equine Genetic Guide

TABLE OF CONTENTS

3   Black
4   Red
5   Champagne
7   Cream
9   Dun
11 Silver
12 Bay
13 Grey
14 Brown and Flaxen
15 Sooty and Mealy
16 Leopard Complex
18 Roan
19 Frame, Tovero and Tobiano
20 Sabino, Splashed White and Rabicano
21 Reorganizing Genese
22 Facial Markings
23 Leg Markings
24 Understanding Basic Genetics
26 Standard Genetic Nomenclature
27 Applying Genetics
32 Genetic Inherited Disorders
34 Genetic Studies

Color Chart at the end of the book for quick reference

The Horse Rainbow: An Equine Genetic Guide

COAT COLORS

Definitions:
The **Points** of a horse are the muzzle, tips of ears, mane, tail and legs from the knees or hocks down.

An **Ermine** is a small black marking located around the coronet and downwards onto the hoof.

A **brand** is an identification mark usually on the horse's shoulder, saddle area or hindquarters.

Freeze branding is done using extreme cold on the hair. Sometime horses are **tattooed** on their upper lip.

Whorls are circular patterns off odd laying hair, like a cowlick, that lay either clockwise or counter-clockwise.

BASE COLORS
There are 16 factors that work together to create coat colors and patterns. They are based on two colors. Every horse has a **Black** or **Red** gene.

Black or (E)
A black horse has all black hairs, not including any markings. Some breeds are only black like the Friesian (which has rare chestnuts too) and some breeds have bred the black out, such as the Haflinger.

Non-fading Black horses do not fade in the sun and the weather. Sometimes it is called Blue Black or Raven.

Fading Black horses do fade from the sun and the weather. It is also called Coal Black.

Foals are born smoky, ashen, brown or bay, and may have lighter legs. Black horses cannot produce Bay foals unless the other parent is Bay or Chestnut.

The Horse Rainbow: An Equine Genetic Guide

Red or (ee)

A Chestnut horse has red hairs that can vary in shades and lightness. Two Chestnut horses will produce a Chestnut. The Silver gene cannot dilute a Chestnut.

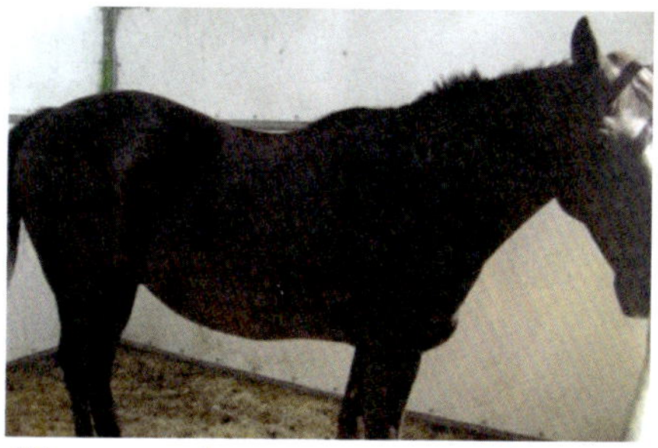

Chestnut and Sorrel mean the same thing genetically. Their mane and tails may be very light or almost white, called Flaxen, to very dark, almost black, but never truly black. Chestnut horses cannot have Black coloring, but they can carry it. If bred to a Black based mate, they can produce Black offspring. There are different shades of Chestnut. The darkest shade is **Black Chestnut** (Black Liver Chestnut). A little lighter is the **Liver Chestnut** (Dark Chestnut).

Brown Chestnut

Standard Chestnut (Chestnut) is a medium shade.

Deeper red shades are called **Red Chestnut** (Red Liver Chestnut) or (Dark Sorrel).

Lighter shades are called **Light, Bright** but often referred to as Gold or Orange Chestnut. Lighter shades of Chestnut are usually referred to as **Sorrel** and have prefixes such as: Blonde, Light, Dusty, Sandy, Clear or Pale. Even lighter are called **Pale Chestnut** and often named Dusty, Blonde or Clear Chestnut.

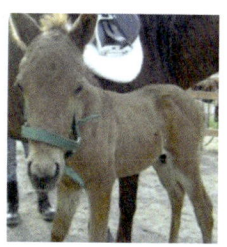

The Horse Rainbow: An Equine Genetic Guide

COLOR DILUTIONS

Diluted coat colors happen when a base coat and a dilution gene are combined. The dilution gene is dominant.

Champagne foals may be a chocolate color.

Champagne or (Ch)

The Champagne gene can dilute both the Black or Chestnut base colors. In order for the coat to be a champagne color, at least one of the parents has to have the color.

A Chestnut horse with the Champagne gene is called **Gold**. Gold horses can have gold or flaxen mane and tails. These horses are often mistaken for Palomino.

Amber is a Bay horse with the Champagne gene. Commonly they are mistaken for Buckskin or Silver Buckskin.

Classic (Lilac) Champagne is a Black coat diluted by the Champagne gene. Often, they are called Grullo, but they are not a Dun dilution.

Sable Champagne is a champagne gene on a brown color. Champagne foals are born dark and their coat lightens with age. They are born with blue eyes that change to green, then hazel, and finally amber. However, some Champagnes keep the blue or green color. They are born with pink skin that will darken to a purple brown. Often, they have **freckles**, especially around the eyes, muzzle, under the tail, and on the sheath or udder areas. Freckling is not the same as mottled or de-pigmentation in the skin. Champagne coats may have a **metallic glow**, which can happen in other coat colors as well. Sometimes the Champagne coat can have **reverse dappling**.

Ivory coats are the color of ivory or have the cream dilution gene. A true **Ivory** (Classic Ivory or Classic Crème) horse's eyes will change color.

Eye changes from 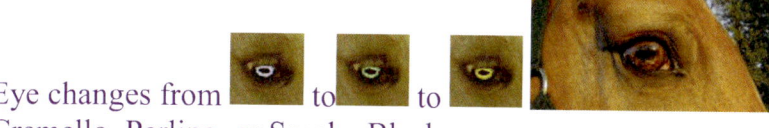 .Often they are mistaken for Cremello, Perlino, or Smoky Black.

When mixed with Crème colors, new colors develop and the horses have pink skin. The brownest is referred to as **Sable Crème.**

A shade lighter is the **Gold Crème** which is on a red base.

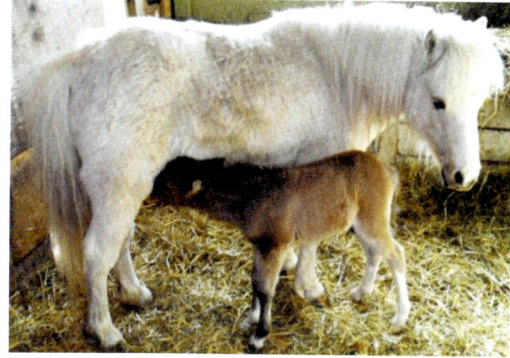

Ivory Crème (Gold Ivory, Amber Ivory and Amber Crème) has a slightly darker yellow hue.

Ivory (Classic Ivory or Classic Crème) is the lighter of the Champagne colors.

The Horse Rainbow: An Equine Genetic Guide

Cream or (Cr)

The Cream gene lightens the base coat. Two cream genes create a lighter the coat. Often the double dilutes are mistaken for Albino, but **Albino** coats lack pigment in hair, skin or eyes. The gene that causes Albinos to happen has not yet been discovered. Buckskins, Smoky Blacks and Palominos with only one Cream gene pass the Cream gene on only 50% of the time.

They are called Single Dilutes.

Palomino coats can be very light, almost white, or very dark and mistaken for Chestnut. They are born very light with pink skin that darkens with age. Sometimes a tuft of hair may be pink. They turn more gold as they mature. The coat color is caused by the Chestnut base color being diluted by the Cream gene. Their manes and tails are light.

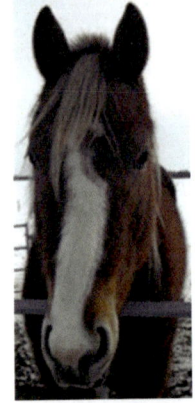  

Chocolate (brown base)   **Caramel Palomino**   (red bases)   **Palomino**

Isabella Palomino (red base)

If a Chestnut base has a cross of the Cream gene and the Dun gene, it is called **Dunalino**. Their coats have primitive markings. The mane and tail are still light.

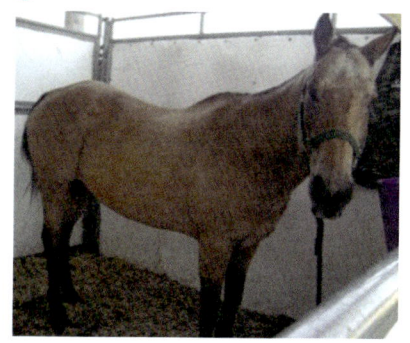

Light and **Dark Buckskin** horses have the Cream gene mixed with the Bay color. They can range from light tan to almost black. They also have black or chocolate points. White guard hairs may be at the top of the tail and the mane.

The Cream gene mixed with the Dun and the Bay genes can create the **Dunskin**, which is just a little darker, especially in the mane and tail, compared to the Dunalino.

Smoky Black has a Cream gene imposed on the Black base. They look like Black horses, but may fade to Brown or Bay.

Double Dilute colors are Cremello, Perlino and Smoky Cream.

Cremello is two Cream genes on a Chestnut base. They always have pink skin and blue eyes. They range from being very light to darker cream. Sometimes white markings can hardly be seen on the very light Cremello coat.

Perlino is double Cream genes on a Bay coat. They usually have darker points, blue eyes and pink skin.

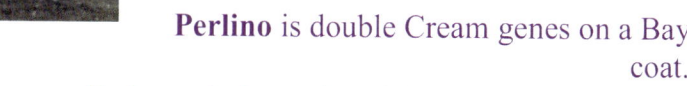

Smoky Cream is two Cream genes combined on a Black coat. One Cream gene will not affect the Black base, but two will.

The Horse Rainbow: An Equine Genetic Guide

Dun or (D)

The Dun gene affects both the Black and the Chestnut base colors. Duns only happen if at least one parent has the Dun gene. The Dun gene causes primitive markings, which are darker than the coat color and can be a dorsal stripe, zebra stripes, shoulder stripes, or cob-webbing. A **dorsal stripe** is a dark line from the base of the middle of the mane to the base of the tail. It is also called an **eel or jack stripe**, a **list**, a **line-back,** or a **back-stripe**. Counter-shading can be confused for a dorsal stripe. Sometimes, a foal is born with a dorsal stripe or other primitive markings or they develop them later on, but they are not Duns. This is considered a camouflage. It is similar to fawns or baby lions with spots that fade over time. Charles Darwin referred to this coloring as an Analogous Variance, meaning that during evolution a trait is lost, but once in a while a baby may be born with it, thus linking us back in time. **Zebra stripes**, **Tiger Stripes** or **bars** are horizontal marks on the legs. They can be dark or light. They fade as they go up the leg. **Shoulder Stripes**, whether dark or light, cross the withers onto the shoulders. **Cob-webbing** or **Spider-webbing** is darker rings or stripes on the forehead. Common eye colors on duns are hazel, amber and grey.

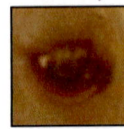 hazel  amber grey

Bay Dun (Zebra Dun) is the Dun gene on the Bay color. They vary in shades of dark to light and of tan to yellow, but usually the points are still dark. Whereas the head of Buckskin will be the same as its body color, a Bay Dun will have a darker head or face mask. A grey shade is called **Silver Dun** and is most noted for the black and white mix of mail and tail hairs.

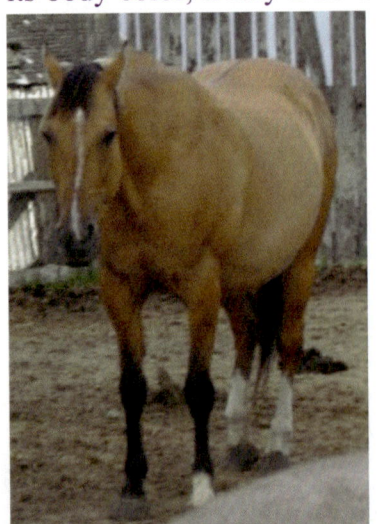

More yellow shades are called **Yellow Dun** or Dusty, Gold, or Buttermilk Dun.

A greyish shade is called **Coyote Dun.**

The Horse Rainbow: An Equine Genetic Guide

Mixing with the Crème gene can lead to the **Dunskin** *(picture in the Cream section)*.

Or it can lead to the **Lilac Dun,** otherwise known as Lavender Dun, which is a pleasant looking mix of the agouti and the crème genes.

Chestnut Dun or **Red Dun** is the Dun gene on the Chestnut base. They do not have black points. Instead the primitive markings are a shade darker red. They range from dark to light. Sometimes they are called Claybank, Red, Chestnut, Apricot, or Peach Dun.

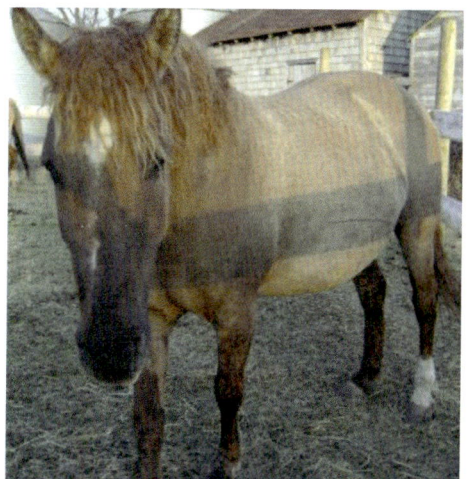

Peanut Butter Dun is Brown Dun.

When mixed with the Crème gene the **Dunalino**, Line-backed Palomino, Dun Palomino, or Palomino Dun are created.

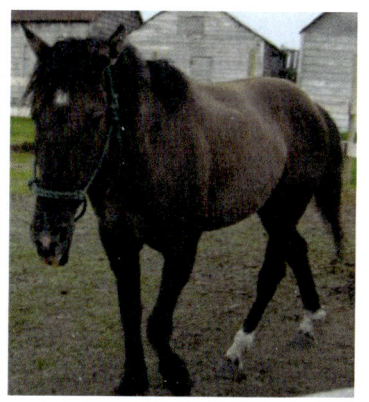

Grullo (Spanish) for males, **Grulla** for females is the Dun gene on the Black base coat. They are smoky or bluish in color ranging from dark to light. They have black points and a darker head. Primitive marks are usually black. Other names are **Slate, Blue** or Black Grullo and the lighter shades are **Light Grullo**, Silver Grullo or Silver Dun, but Silver is misleading due to the color of what the Silver gene can actually do to the coat.

When mixed with Brown it creates the **Olive Dun** or Mouse, Muddy, Wolf, Dove, or Lobo Dun.

The Horse Rainbow: An Equine Genetic Guide

Silver or (Z)

The Silver gene only affects the Black base coats. They are **Silver Dapple** (*left*), or Chocolate Blue Silver by the Mountain breeders, or Black or Blue Silver by the Icelandic breeders. Others may call it Grey but it is incorrect due to the lack of color change as the horse ages. They range in shade. The darkest dilution on a Black base coat is a light to **Chocolate**  **Silver** (*right*) with a mane and tail that ranges from a sooty slate color to silver white. Sometimes, they are confused with Chestnuts. They do have dark points; however, light hairs are mixed in.

Silver Bay is also **Red Chocolate Silver**, which is the Silver gene on a Bay color. The bodies are not affected, but the points are lighter or sootier colored. The mane and tail, whether light or dark, will look sooty. Foals have light legs.

Silver Buckskin has a gold body and diluted points. They have sooty, darker legs, dark ear tips and the mane will vary from light to dark, but will be silver and sooty.

COLOR MODIFIERS, ADDITIVES & RESTRICTIONS

Modifying genes are either subtle or extreme.

Bay (Agouti Gene) or ((A)

They have a red body and black points. The **Bay** color is a Black based color that has been modified. The shades are called:
Black Bay (Seal Brown or Seal Bay)

Dark Bay (Standard or Brown Bay)

Burgundy Bay or Mahogany Bay is almost a rich purple/red color.

The most common bay is a **Red Bay** (Blood bay or Cherry Bay)

A lighter shade is a **Bright Bay** or Copper Bay.

Lighter yet can be confused with buckskin, but there is no crème dilution and there is a little more red in the coat called **Light Bay** (Sandy Bay or Golden Bay).

Wild Bay is a bit different. The black markings on the legs extend only up the pastern and fetlock, then it fades out.

The Horse Rainbow: An Equine Genetic Guide

Grey or (G)

Grey is a modifier of the base color and it slowly removes the color, called de-pigmentation. The Grey gene can mask all other colors and all horses carrying the Grey gene will end up a shade of Grey or White. It is a strong gene. A Grey horse has at least one Grey parent. If both parents were Grey, the offspring will always be Grey. De-pigmentation on the skin is when the skin loses its color over time, similar to the Appaloosa pattern. Sometimes it is called **Pinkie** or **Fading Arab Syndrome**, but medically it is called **Equine Vitiligo**. It can also occur in other coat colors.

Grays get lighter with age. The first sign is usually **goggles**, which is graying around the eyes. Grey hairs are more likely to occur around the eyes, ears, muzzle, mane, or tail.

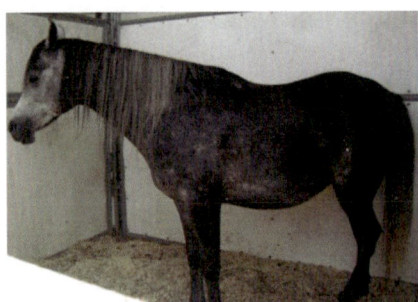  

Steel Grey or Iron Grey colored horses are at the beginning of the graying process. They have a blue tint. Next is **Medium Grey**. **White Grey** (Light Grey) is the final product of the process. Most of the pigment has been removed and just the muzzle remains dark.

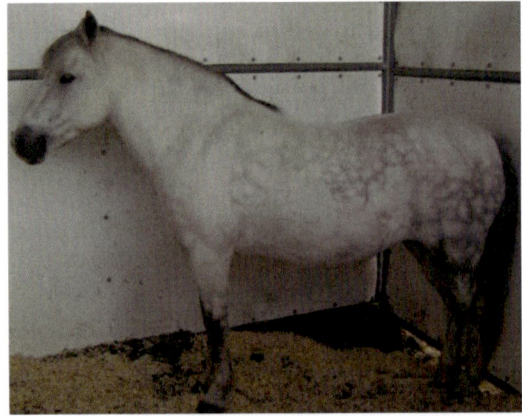

Dapple Grey has blotches of white gray markings.

Flea-bitten Grey horses have small red or black dots on the coat. These will lighten with age as well. It is not a sign of what the base color is.

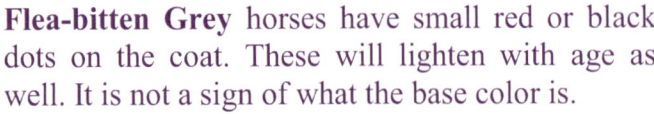

The Horse Rainbow: An Equine Genetic Guide

 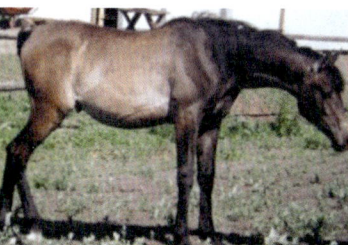 

Dark Rose Grey (Burgundy Rose Grey or Burgundy Grey) and Rose Grey (Peach Rose Grey or Light Rose Grey) horses have a Bay coloring mixed with grey, but are not roan. **Mulberry Grey** or (Chestnut Grey) horses have a unique mix of brown, black and white hairs.

Tetrarch Spots are unusual markings.

Blood Marks or Bloody Shoulder is very rare and usually happens in Arabians. It is a large concentration of fleabites in one area.

<u>**Brown is undefined**</u>
Not much is known about how the solid color of brown without black points comes into play in horse color coats.

<u>**Flaxen ((ff)**</u>

The **Flaxen** gene affects the mane and tail, causing them to be white or cream. Not much is scientifically known about this gene.

The Horse Rainbow: An Equine Genetic Guide

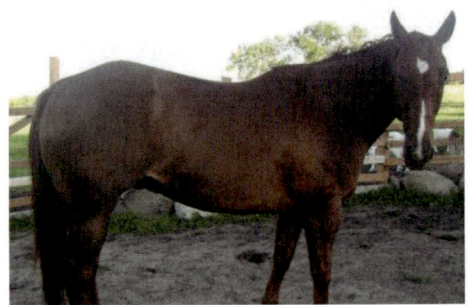

Sooty or Smutty or (Sty)
The gene acts on either Chestnut or Black bases by darkening the color or certain parts. It can cause dapples, or black hairs. **Sooty** looks darker on the top of the horse in a dark line and dark head, called **Counter-shading**. The parts least affected are the **soft areas**: under the belly flanks, behind the elbows, buttock, muzzle and eyes.

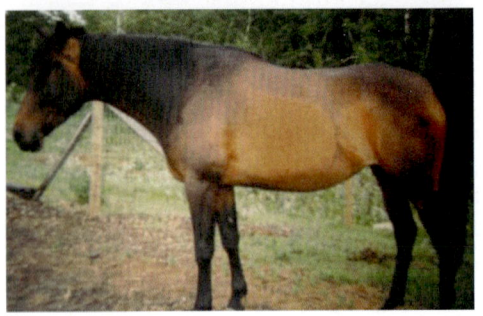

Sooty on:

Chestnut Bay Buckskin
Palomino

Mealy or Pangare or (P)
The Mealy or Pangare gene modifies the coat color in the soft areas. Horse colors usually range from light to tan. **Mealy Mouth** is another name.

The Horse Rainbow: An Equine Genetic Guide

WHITE PATTERNS

White markings or patterns are always colored over top of the horses coat color. The Roan pattern happens when the white hairs intermix with the colored hairs. The Appaloosa or Paint patterns happen when the white hairs make up solid white patches of hair. A **White or (W)** horse has still has pigment. No pigment is **Albino** or **(ww).**

Leopard Complex or (Lp)
The **Appaloosa** pattern or **Leopard Complex** and **Semi-Leopard** do not affect the color of the horse. These white markings are put on over top of the color. This pattern can cause the mane and tails to be sparse, or **rat-tailed**. The white can be minimal or have maximum coverage. 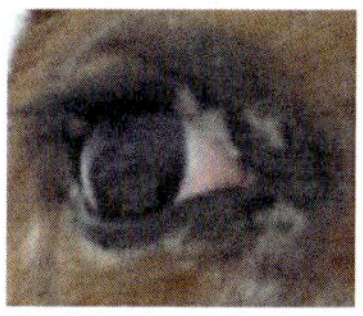
Mottled Skin is small pink or white dots around the eyelids, mouth, genitalia or anus.
White Sclera of the eye is typical. **Striped Hooves** happen regardless of the leg color.

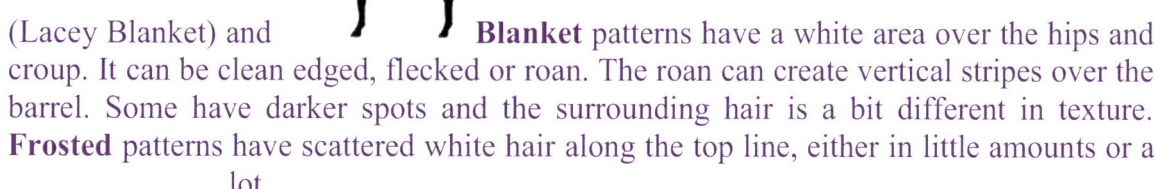

Extended Blanket, (Lacey Blanket) and **Frosted Blanket**

Blanket patterns have a white area over the hips and croup. It can be clean edged, flecked or roan. The roan can create vertical stripes over the barrel. Some have darker spots and the surrounding hair is a bit different in texture. **Frosted** patterns have scattered white hair along the top line, either in little amounts or a lot.

 **Few-Spotted Leopard** patterns happen when the white covers most of the horse, with a few spots remaining colored. **Hip-Spots** are dark spots on white.

 **Minimal Blanket** and  **Loin Blanket**

The Horse Rainbow: An Equine Genetic Guide

Leopard horses are white with spots.

 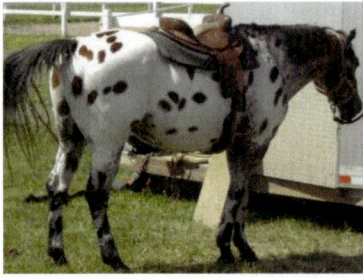

Semi-Leopard

Snowcap patterns happen when the blanket is all white.

Snowflake patterns have small spots randomly over the body, often mistaken for Bird-catcher Spots.

Varnish Roan is not graying or classic roan, but rather a colored horse that gets lighter with age.

Pintaloosa (mixed patterns)

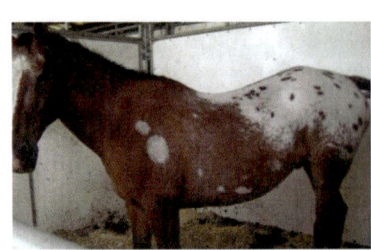 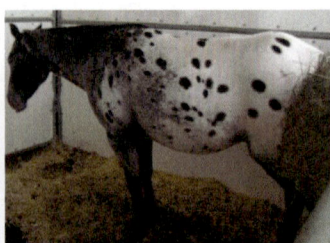 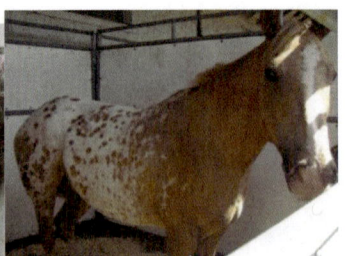

The Horse Rainbow: An Equine Genetic Guide

Roan or (Rn)

Roan happens when white hairs mix with colored hairs. Roans have darker points. **Classic Roans** do not have white hairs on the head, mane and tail or lower legs.

Flowery names can be attached to this pattern, such as Purple Roan, Honey Roan, or Lilac Roan. Roans are less common than Grays and Roans are not progressive, but they may get darker. Heads, legs, mane and tails will be solid colors, with maybe a few stray white hairs. Injuries heal over the base coat hair coloring. At least one parent needs to be a Roan for a foal to be Roan. Two Roans may produce a Homozygous Roan foal, which may be lethal, but it is not scientifically proven.

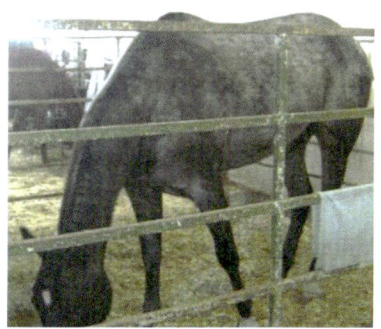

Black Roan

and **Blue Roan** happens on a Black base coat.

It causes a Grey, Blue, or Purple tint.

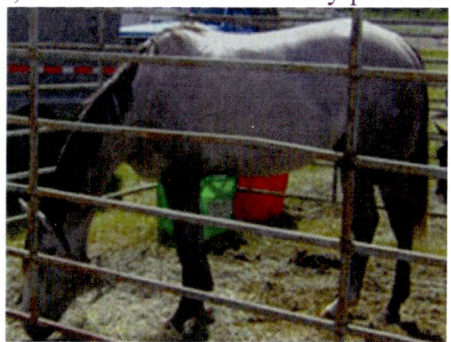

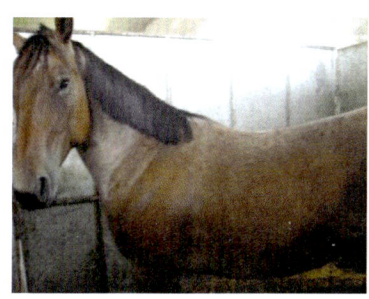

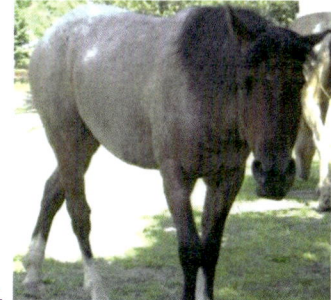

Bay Roan is roan on a Bay color.

Chestnut Roan, Red Roan, or Strawberry Roan is roan on a Chestnut base.

There are also **Palomino Roan** and **Buckskin Roan**.

Frame is (Fr) also is Overo or (O)

Skewbald is a term for a solid color and white pattern. **Piebald** is the same for a black colored horse.

Frame is the **Overo** pattern. A frame of color surrounds the horse. They have dark legs and may or may not have

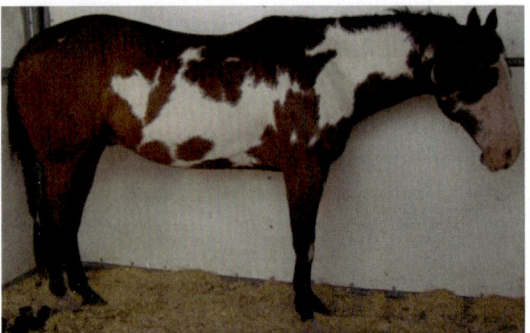

one or two blue eyes even if color surrounds the eye. The white pattern rarely touches to top-line and usually has cleaner lines, but jagged. Horses with a white facial marking have a pigmented upper lip called a **mustache**. If the eye is blue or pale white blue it is called a **walleye**. Two Frame horses produce a foal with **Overo Lethal White Syndrome**, a lethal production of a white baby with malformations.

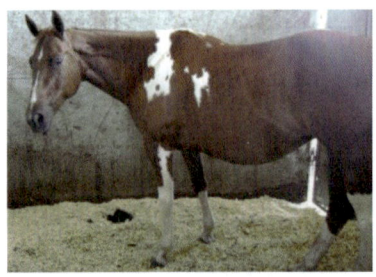

Mixed Patterns
This pattern happens when the Leopard Complex pattern is mixed with any of the Frame, Sabino, Tobiano, or Splashed White pattern.

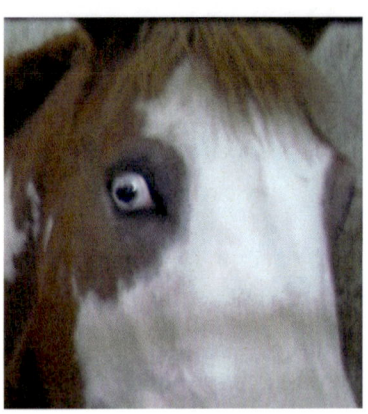

Tovero is the Tobiano pattern mixed any of the three Overo patterns.

Tobiano or (To)

Tobianos have solid colored heads. White on the face means that the horse is Frame, Sabino, or Splash. A Tobiano with a blue eye may carry the Frame gene and when crossed, may produce the OLWS foal (lethal white). The white pattern is more vertical looking and will cross the top-line of the horse. The spots will have clean edges and be more rounded. The tail

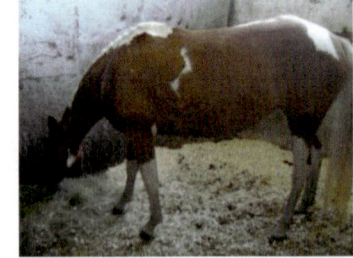

will be white or have a bit of color at the bottom. One pattern is spotting on the flanks and chest, called **shields**. Small spots are called **Ink Spots, Paw Prints, Cat Tracks,** and so on. Some Tobianos have pigmented skin that travels into the white area, resulting in a shadow or halo. These horses are called **Shadow Paints**, **Ghost Paints**, **Ribbon Paints**,

The Horse Rainbow: An Equine Genetic Guide

or **Halo Paints**. **Ermine spots** or **Distal Leg Spots** are amounts of color in the white of the leg. Sometimes Tobianos do not have any body spots so they are called **Slipped Tobianos** or **Breeding Stock Paints**.

Sabino or (Sb)

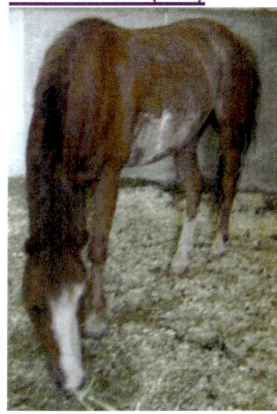

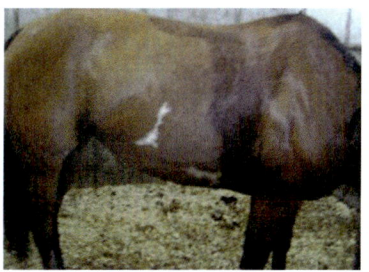

Sabinos is another Overo term, meaning to have large facial markings, or chin spots. Their legs are mostly white, or at least above the knees. Sometimes Sabinos have blue eyes. A little bit of Sabino expressing itself causes Roan. They may not have Sabino coloring at birth, but it will develop over time. Horses are said to have *chrome* if they have lots of white on the face and lots of white on the legs. The white can express itself anywhere on the horse.

Splashed White or (Spl)

Splashed White is another Overo pattern, where white patterns on the face are bottom heavy. Leg whites can vary, but do not travel above the knee. Belly spots are common. If the face has lots of white, the horse is usually a "bald-faced" with just the ears being colored.

Rules for Overo: Body 1 color, withers to tail colored, 1+ colored legs, irregular white, mostly white head, 1 color tail
Rules for Tobiano: Solid head with markings, colored flank(s), 4 stockings or legs, smooth white, 2 color tail
Rules for Tovero: Dark ears and mouth, Blue eye(s), colored chest, flanks and tail spots

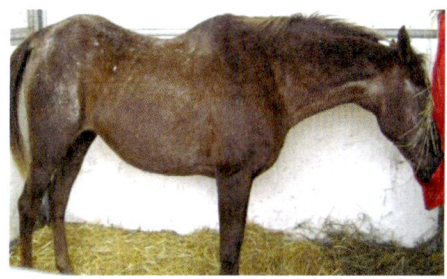

Rabicano or (Rb)

Rabicano happens at the base of the tail, flank and belly. It can be subtle or quite clear to see. Some horses may have a large white spot on the underbelly or between their hind legs. It can also be called ticking, highlights, or skunk tail.

REORGANIZING GENES

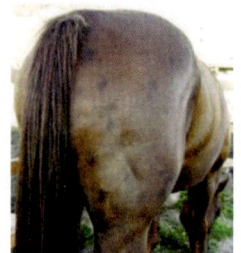

Bends or Spots - Bends or Spots are also called Ben d'Or, Smuts, or Grease Spots. They are darker spots located on the body.

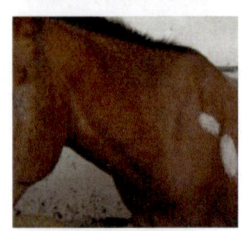

Bird-catcher Spots - Bird-catcher Spots are small white dots on the body. They may be linked with the Sabino gene.

Brindle - The Brindle pattern is very rare. Usually the main body (neck, shoulders, quarters) are striped. Some Grey horses may show Brindle patterns due to the Graying process.

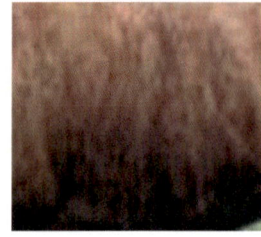

Lacing - Lacing can be called Cobweb or Lace. Not much is known about this pattern. Some say is because of a fungal infection or a blood disorders. Some say it's linked to the Appaloosa or Paint patterns. Nothing is scientifically known for sure. It is sometimes referred to as giraffe markings.

Marble - Marble is a unique pattern of white or light colored hairs creating an effect similar to the pattern found on giraffes. It can also be called Giraffe markings. It is very similar to Lacing but has some darker areas mixed in.

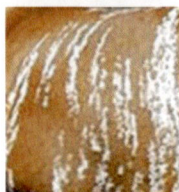

Camouflaging - Camouflaging is dark colored stripes appearing on the horse, creating a zebra like effect. They usually appear in certain spots only. Sometimes they will cross the shoulder, neck, rump and legs.

White Striping - White stripes are the opposite look of camouflaging and can appear on the colored horse. It creates an effect of white zebra stripes. They usually appear in certain spots only. Sometimes they will cross the shoulder, neck, rump and legs.

Tan Markings - Occasionally, a horse may have tan colored markings instead of white markings.

The Horse Rainbow: An Equine Genetic Guide

22
FACIAL MARKINGS

The face may have a number of markings

Star- on the forehead

Stripe or **Strip**- narrow white line

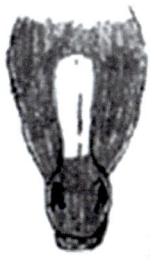

Blaze- wide white line that extends over the nasal bones

Snip- between or into the nostrils and below

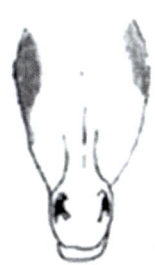

White face or **Bald-faced**- includes forehead, eyes, nose parts of muzzle

The Horse Rainbow: An Equine Genetic Guide

LEG MARKINGS

Ermine- black marks on the coronet band and downwards onto the hoof

Coronet- around the coronary band

Pastern- extending over the heel and below the fetlock joint

Fetlock- touching the fetlock joint, but not above

Sock- above the fetlock joint, but less than half way to the knee

Half-Stocking- more than half-way to the knee, but not touching the knee joint

Stocking- to the knee joint

Leg- above the knee

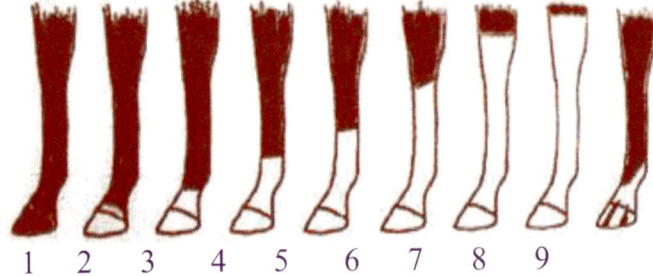

1.	None/Dark hoof
2.	Coronet
3.	Pastern
4.	Fetlock
5.	Sock
6.	1/2 Stocking
7.	Stocking
8.	Leg
9.	Ermine

Unique disconnected marking

HOOF MARKINGS

White- an all-white hoof

Dark- all dark hoof

Parti-colored- a white hoof with a dark stripe, called an Ermine

UNDERSTANDING BASIC GENETICS

Genetics is the science of heredity, the molecular nature of genes and the biochemical reactions with them.

Genes make up the horse's genotype, in which the genetic qualities are inherited and passed down from generation to generation.

Phenotype applies to how the horse looks physically or a physical feature.

Alleles are an alternative form of gene, or gene mutations, found at the same place as a chromosome. When looking at a genetic formula the capital letter is dominant over the small letter which is recessive. A co-dominant allele creates the appearance of both physical traits.

Chromosome is a paired gene, which horses have 64 chromosomes located in the nucleus.

DNA strands consist of segments of chromosomes.

A strand of DNA, note the chromosome pairs

A particular place on a chromosome is a **locus**, which a particular gene can always be found. **Gametes** or sex cells carry half of the information from each parent, which information it carries is determined by chance.

When the egg cell fuses with a sperm cell a **zygote** is formed. Through a process called meiosis, cells divide to form a new gamete cell, each carrying one copy of the parent's chromosomes. Chromosomes may also **cross-over**, but still retain the same information, but the new chromosomes have been recombined. The end result is a genetic variation which forms inherited diversity.

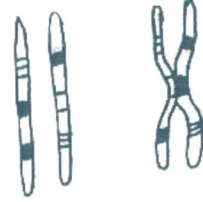

Chromosomes in pairs, either remain the same or sometimes connect, crossing-over

Homozygous means the genes from both parents are the same in a particular gene combination.

Heterozygous is when they are different.

MOLECULAR GENETICS & BIOCHEMISTRY

Melanin is the pigment that creates color. Brown-black pigment is called **eumelanin** and red-yellow pigment is called **phaeomelanin**.

Melanism is an increased amount of black or nearly black pigment.

Abundism is an increase in dark pigmentation, creating stripes and patches.
Overlapping of abundism is **pseudo-melanism**.
Melanism happens because of changes in the agouti gene, controlling banding of black and light areas on the hair shaft.
Melanism may also play a role in immunity, as there is a correlation with resistance to viral infections and epidemics. This may be due to a protein on the cell membrane, as well as due to an increased absorption of heat.

Under a microscope on low power a black hair can be distinguished from a lighter hair

STANDARD GENETIC NOMENCLATURE

Genetic Notation or Shorthand Letters used to identify genes and alleles.

All horses have either a red or black base coat. All other colors are built up on top of the base coat. Genes string together to create different colors. The EE or Ee genes cannot produce chestnuts or sorrels, only if the gene is ee.

Partial List of Genetic Formulas

Black Layering Colors (EE or Ee)

Smokey Black is EE, CcrC, which is black with Cremello allele.

Grullo/Grulla is EE, Dd, which is black with the Dun allele.

Blue Roan is EE Rr, which is black with the Roan allele.

Bay is AA, EE, which is black with the Agouti gene or the bay modifier allele.

Buckskin is AA, EE, Dd, which is bay with the cremello allele.

Dun is AA, EE, Dd, which is bay with the dun allele.

Roan is AA, EE Rr, which is bay with the roan allele.

Red Layering Colors (ee)

Chestnut or Sorrel is ee, and all other genes are recessive except for variation of colors.

Palomino is ee, CcrC, which is chestnut with the Cremello allele.

Red Dun is ee, Dd, which is chestnut with the Dun allele.

Red Roan is ee, Rr, which is chestnut with the Roan allele.

Dunalino is ee, CcrC,Dd, which is chestnut with the cremello and Dun allele.

APPLYING GENTICS

A single black gene will dominate a red gene, therefore all chestnut horses are homozygous red.
Flaxen horses are also homozygous because the flaxen gene is recessive and needs to be the same from both parents. In order for a horse to be gray, one parent needs to be gray.

Notation of Gene Alleles to Color

GeneAlleles Horse Color

E	E	Black pigmented skin and hair either overall or just on points pattern
E	e	Same as above
e	e	Black pigmented skin but coat is red, unless white, gray, roan or dilution genes are present. * + can also be d
E+	E+ or e+	Occurs on a different locus, which overrides Agouti alleles resulting in a dominating all black horse who does not fade
e+	e+	No effect

A	A	Agouti gene restricts eumelanin to points pattern = EE or Ee and A has no effect on red pigment = ee
A	a	Same as above
a	a	If black hair = EE or Ee, then black is uniform overall and no points A has no effect on red pigment = ee

Sty	Sty	Sooty, black mixed in with coat color
Sty	sty	Same as above
sty	sty	No black mixed in

P	**P**	Pangare or Mealy effects alleles located in soft areas of the horse: Muzzle, girth, flanks, buttocks, around eyes and elbows
P	**p**	Pangare signs
p	**p**	No mealy

F	**F**	Flaxen gene affects red only, creating red points
F	**f**	Same as above
f	**f**	Creates Flaxen points on ee

Ch	**Ch**	Champagne is rare dilution, creating bronze glow, pumpkin colored coat with freckles, amber or blue eyes. Skin around eyes is pink with freckles. Born darker but lighten
Ch	**ch**	Champagne signs
Ch	**ch**	No champagne dilution

C	**C**	Full pigmented, Cremello or Perlino. Double dilute, where red pigment is diluted to pale cream and black pigment is diluted to red shade. Skin and eyes may also be diluted, blue eyes are common
C	**c**	Single dilution, red pigment will be diluted to yellow/gold but cream/white mane and tail, but black is unaffected, creating Palomino, Buckskin or Smokey Black
c	**c**	No dilution, horse will have full pigment

D	**D**	Dun color will be diluted to pink-red, yellow-red, yellow or mouse gray and will have dark points, dorsal & shoulder stripe and bars on legs.
D	**d**	Same as above
d	**d**	Undiluted coat

The Horse Rainbow: An Equine Genetic Guide

Z	Z	Silver Dapple, dilutes eumelanin to brown with white mane and tail
Z	z	Same as above
z	z	No silver dapple

G	G	Born non-gray, but grays later, with progressive de-pigment
G	g	Same as above
g	g	Does not de-pigment with age

* O can also be FR

O	O	Frame/ Overo white spot pattern, includes Sabino & Splash
O	o	Same as above. NOTE: Homozygous is Lethal White Syndrome
o	o	No Overo pattern

Sb	Sb	Sabino markings, which is rough white markings, made by a complex gene allele
Sb	sb	Same as above
sb	sb	No Sabino marks

Spl	Spl	Splashed white pattern, white markings upward, blue eyes common
Spl	spl	Same as above
spl	spl	No splashed white markings

TO	TO	Regular & distinct white spot pattern, Tobiano, legs are usually white
TO	to	Same as above
to	to	No Tobiano pattern

Rb	Rb	Rabicano pattern, which is partially like Roan, white in tail
Rb	rb	Same as above
rb	rb	No Rabicano pattern

Rn	Rn	Roan pattern, which is white mixed with base colored hair
Rn	rn	Same as above
rn	rn	No roan pattern

p	Lp	Appaloosa or Leopard Complex, which creates spots.
Lp	lp	Same as above
lp	lp	Allele is minimal or suppressed in E or E+

W	W	Lethal white, incomplete color, inability to defecate (gene near W or O that does not influence the color)
W	w	Horse is white but lacks pigment
w	w	Horse if fully pigmented

The **Mandelian Inheritance Rule** states the possible genotype characteristic of the offspring can be determined by the dominance of the main gene of the parents. We can predict the probable outcome of the foal. Using a **Punnett Square** we can predict the probability of the outcome of a crossing between two horses.

1) Example of Breeding a Gray Mare and a Chestnut Stallion
Detecting Color Ratio Outcome
Mare Egg is 50/50
Stallion Sperm has Only G+ alleles

| G+ | 50% chance of G+G+ or Not Gray |
| Gg | 50% chance of GgG+ or Gray |

Outcome is 1:1 ratio foal will be not Gray to Gray

The Horse Rainbow: An Equine Genetic Guide

2) Example of breeding a heterozygous gray mare and a heterozygous gray stallion.
Detecting Color Ratio Outcome
Mare Egg is 50/50
Stallion Sperm is 50/50

	G+	Gg
G+	25% G+G+ (not gray)	25% G+Gg (gray)
Gg	25% GgG+ (gray)	25% GgGg (gray)

Outcome is 3:1 ratio foal will be Gray to not Gray.

This is a **mono-hybrid cross.**

3) Example of a breeding of two homozygous paints to get a paint to breed to a champagne to create a champagne tobiano.
Breed this champagne tobiano to another.

Mare combinations:
ToToChCh   or   ToToChch   or   TotoChCh   or   TotoChch
Stallion: possible Allele combinations
ToToChCh   or   ToToChch   or   TotoChCh   or   TotoChch

	ToToChCh	ToToChch	TotoChCh	TotoChch
ToToChCh	1	1	1	1
ToToChch	1	1	1	2
TotoChCh	1	1	3	3
TotoChch	1	2	3	4

1 – Champagne Tobiano
2 – Chestnut Tobiano
3 – Golden Champagne
4 – Chestnut

Outcome for Champagne Tobiano to another color is 10:16 or 5:8 or 62.5%. The outcome for a chestnut is very rare only 6.25% chance. A chestnut tobiano would be 18.75%. A golden champagne would be 18.75%.

This is called **Di-Hybrid cross.**
This cross is based solely on simple gene action. In reality probability is more complex and an allele has to be dominant over another one in order to show in the phenotype. Sometimes they can be partially dominant, which creates a medium between the two.

GENETICS AND CO-DOMINANCE
Sometimes alleles are there but not expressed in the phenotype. Because they do not make any alterations affecting breeding selection, they can exist forever. Sometimes these genes are responsible for disorders, such as CID or Equine Combined Immune Deficiency Disease. These genes are recessive, making the offspring need one gene from

each parent in order to contract the disease. Offspring with only one gene will be carriers. Epitasis is where one gene masks another. This may cause horses to be born one color and change to another over time.

Non-sex linked genes are called autosomes.

GENETIC INHERITED DISORDERS

Congenital anomalies and inherited disorders are all physical abnormalities at birth and those diagnosed later.

AUTOIMMUNE SYSTEM
- **Severe Combined Immunodeficiency (SCID)** is a fatal disease of Arabian and part-breds. It is an autosomal recessive trait. Once colostrum antibodies decrease there is no more immunity. Foals develop infections and die.
- **Neonatal Isoerythrolysis** is not inherited, as the absorption of colostrum antibodies destroys red blood cells. The dam must have a certain negative blood type and be sensitized to an offending antigen, such as through other pregnancies, blood transfusions or placental contamination. The sire must also have these antigens. After nursing, foals get sick after nursing, jaundiced and anemic.

EYES AND EARS
- **Colobomas**
- **Absence of the nasal punctum**
- **Blockages** or **drainage parts absent**
- **Entropion eyelids**
- **Cataracts** are all congenital defects.

HEART DEFECTS
- **Holes** or **defects** between the atria or ventricles, or things like **closure failure** of circulatory system

GASTROINTESTINAL SYSTEM
- **Blockages** in the intestinal tract
- **Lethal white disease** is related to the White gene and not limited to breed
- **Brachygnathia** is a malocclusion between the mandible and the maxilla, creating a parrot mouth
- A **cleft palate**, where Aspiration pneumonia is may result

The Horse Rainbow: An Equine Genetic Guide

MUSCULOSKELETAL SYSTEM
- **Limb deformities** may be a result of being too loose, too tight, bow-legged, knock-kneed and so on
- **Patellar luxation**, usually only happens in miniature horses and Shetlands
- **Skeletal malformations** are a twisting of the spine, or sometimes the spine has incomplete closures
- **Hydrocephalus**, which is abnormal fluid accumulation
- **Malformed digits**
- **Malignant hyperthermia syndrome**
- **Hernias**, either **umbilical** or **inguinal**
- **Dwarfism**
- **Tying-Up Syndrome**
- **HYPP or Hyperkalemic periodic paralysis**

RESPIRATORY
- **Guttural pouch tympany**
- **Choanal atresia**

SEX DETERMINATION
- The intersex disorders **hermaphrodite** and **pseudohermaphrodite** is a mixture of sex hormones

SKIN
- **Junctional mechanobullous**, is a hoof problem
- Undermined skin or **Hyperelastosis cutis**
- **Dilute Lethal** or **Lavender Foal Syndrome** occurs in Egyptian and part-bred Egyptian Arabians

UROGENITAL TRACT
- **Ruptured bladder**
- **Patent Urachus**, urine leaks from the umbilicus
- **Recto-vaginal, urethro-rectal fistulas**
- **Ectopic ureter**, although is very rare
- **Uterus unicornis** is a single uterine horn not two

BIOCHEMICAL GENETICS & BLOOD GROUPS
Biochemical genetic markers prove parentage. Blood group systems are analyzed for variation of using antibodies and standard serological techniques. Genetic serums or red cell proteins or enzymes are detected. Genetic analysis provides information about variations. Information can predict how breed selection will influence genetic variation.

IMMUNOGENETICS & GENETIC ASPECTS OF DISEASE

Genetics in relationship to immunity responses can show how and why some horses are immune to diseases and sickness, in order to help better understand the disease to perhaps develop a cure or a vaccine. Immunoglobulin genes, light and heavy, have distinct functions in the immune system, which play a vital role in the health and breed selection. This can be used to develop better vaccines and methods to improve the immune system. Specific breeds may be resilient or susceptive to certain disease. Sometimes it is hard to tell whether genetics or environment influences disease.

GENETIC STUDIES:

CYTOGENETICS AND PHYSICAL CHROMOSOME MAPS

Cytogenetic studies the structure of chromosome material. The karyotype of the horse consists of 64 chromosomes, where each can be identified by band-staining techniques which allow recognition of minor structural rearrangements of chromatin material.

Primary infertility in mares, usually happens to chromosomal abnormalities resulting in inactive ovaries. Secondly, infertile mares may have inactive gonads, associated with chromosomal rearrangements called balanced reciprocal translocations.

Small, unthrifty foals may have abnormalities of the autosomes, not the sex chromosomes. An extra chromosome is recognized in these cases.

GENETIC LINKAGE MAPS

A gene map shows complex hereditary basis of behavior, performance and disease, allowing information to be spread to help prevent death and disease. This will foster better breeding selection and create better welfare for horses.

Studies need to show genetic markers on the horse genome, which has 60,000 to 80,000 genes-pieces of DNA codes. Genetic variations show which type is better or worse when compared. New genetic systems are being recognized all the time. Genetic basis allows scientists to map genes to isolate major genes responsible for this disease.

DEVELOPMENTAL GENETICS

Many fear the 'Super-horse' clone. Genetic traits can lead people to clone horses of superiority, leaving chance out of the breeding selection. However an artificially reproduced horse still has many questions to answer. Currently, clones have problems with premature aging, bone problems, obesity and decreased immunity. Offspring will look normal but develop severe problems later on. There is no research done with brain activity and thinking patterns. While cloning presents a very interesting role in life and science it will not replace the chance of breeding naturally to gain new super horses.

To clone, an egg is harvested then emptied of all of its DNA material and replaced with the gene material of the horse to be cloned. The egg is planted in a female and hopefully conception will happen.

There are also questions of what cloning will do to the breeding market. People will want to choose super-horse clones for their mares, leaving new stallions alone and small studs will be forced to close their operations. The cost of cloning is still quite expensive.

GENETIC RESOURCES AND CONSERVATION

Knowing a genetic bank of information will help industrial biotechnology. This includes a Genetic analyses, DNA monitoring, storage and cryoconservation. It will help preserve genetic stock. This is done to help keep certain genetic traits in the world, in hopes that they will benefit future generations. Because populations have derived from random mating, gene pools can be diverse. However they are restricted by geography, finance, appearance and so on. Genetics also interacts with nutrition, exercise, environment and medical care. Breed groups can confine inherited genes through the ages. Genes will be untouched by science. This also leads to the problem of inbreeding, which can help or hinder the gene pool.

GENETICS OF PERFORMANCE TRAITS

Considered traits can range from anything. Racing traits are time and final rank, which affects breeding selection. Population genomics identifies chromosome affecting which genes are harbored. This affects morphological and physiological traits important for athletic performance.

However, these genes can still vary. Nevertheless, the breeders will still want to select the finest of the species and try to better the species.

The Horse Rainbow: An Equine Genetic Guide

BASE COLORS: BLACK (Must be all black)

 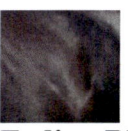   

Non-Fading Black **Fading Black** **Black Foal** <u>Undefined</u> <u>(ff) Recessive</u>
Raven *Coal Black*   **Brown** **Flaxen**
Blue Black

RED or CHESTNUT

Black **Liver** **Brown** **Standard** **Red** **Light** **Pale/Sorrel**
Black Liver *Dark*   *Red Liver* *Bright Red* *Dusty*
            Dark Sorrel *Gold* *Blonde*
               Orange *Clear*
            Blonde Sorrel *Sandy Sorrel*
            Sorrel

MODIFIERS:
BAY (Red with black points)

Black **Dark** **Burgundy** **Standard** **Bright** **Light** **Wild Bay**
Seal Brown *Brown Bay* *Mahogany Bay* *Cherry Bay* *Copper Bay* *Sandy Bay* *Any color but with*
Seal Bay   *Blood Bay*   *less black points*
            Golden Bay
            Red Bay

GREY (Progressive graying with age)

   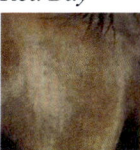 

Steel Grey **Medium Grey** **White Grey** **Dark Rose Grey** **Peach Rose Grey**
Iron Grey   *Light Grey* *Burgundy Rose Grey* *Light Rose Grey*
         Burgundy Grey

Mulberry Grey **Flea-bitten** **Tetarch** **Blood-Marks** **Dappled**
Chestnut Grey   *Bloody Shoulder*

SOOTY or Smutty   PANGARE or MEALY

Chestnut **Bay** **Buckskin** **Palomino**   *Any color with mealy pattern*
            Mealy ... or M*ealy Mouth ...*

DILUTIONS:

CHAMPAGNE (Metallic glow, reverse dappling)

Classic **Sable** **Gold** **Amber** **Sable Crème** **Gold Crème** **Ivory/Classic Ivory**
Lilac *Classic Crème*

Eye changes from to  to
  Blue Green Amber

Ivory Crème **Reverse Dappling**
Gold Ivory
Amber Ivory
Amber Crème

Champagne Foal

CRÈME (single dilution)

Smoky Black **Dark Buckskin** **Light Buckskin** **Chocolate Palomino** **Caramel Palomino** **Palomino** **Isabella Palomino**
Light Palomino

(double dilution)

Eye colors

Smoky Crème **Cremello** **Perlino**
  Hazel Amber Grey

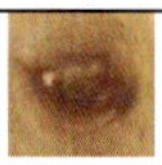

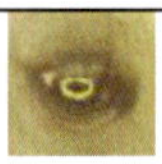

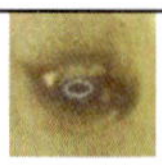

Common to have **Hazel eyes**

DUN (Must have dorsal stripe)

Dark Grullo **Light Grullo** **Olive Dun** **Coyote Dun** **Bay Dun** **Yellow Dun** **Silver Dun**
Slate Grullo *Silver Grullo* *Mouse or Muddy* *Zebra Dun* *Dusty or Gold* *Oatmeal Dun*
Blue Grullo *Wolf or Dove* *Buttermilk or* *White Dun*
Black Grullo *Or Lobo Dun* *Pale Dun*

Dunalino **Dunskin** **Lilac Dun** **Peanut Butter Dun** **Red/Claybank Dun**
Line-backed Palomino *Line-backed Crème* *Lavender Dun* *Brown Dun* *Chestnu/ Apricot or Peach*
Dun Palomino *Dun*
Palomino Dun

SILVER (Mane and tail are light and dark)

Black Silver   **Chocolate Silver**   **Silver Bay**   **Silver Buckskin**   **Silver Dapple**
Blue Silver   *Red Chocolate Silver*   *Chocolate Blue Silver*

ROAN (One or two colors of hair intermingle with white hairs)

Black Roan   **Blue Roan**   **Red Roan**   **Strawberry Roan**   **Purple Roan**   **Bay Roan**   **Roan Palomino**
Grey Roan   *Black Blue Roan*   *Chestnut Roan*   *Lilac/Lavender Roan*

Buckskin Roan   **Classic Pattern**   **Corn Pattern**   <u>RABICANO</u>   **Rabicano**
Roan not on face   *Solid patches*   *Skunk Tail*

LEOPARD COMPLEX (Mottled skin)

Leopard   **Semi-Leopard**   **Extended Blanket**   **Blanket**   **Frosted/Lacey Blanket**   **Few Spotted**

Hip Spots   **Snowflake**   **Snowcap**   **Minimal Blanket**   **Loin Blanket**   **Pintaloosa**   **Varnish Roan**
                                                                                  (Mixed)

OVERO or FRAME

TOVERO (mixed with Tobiano)

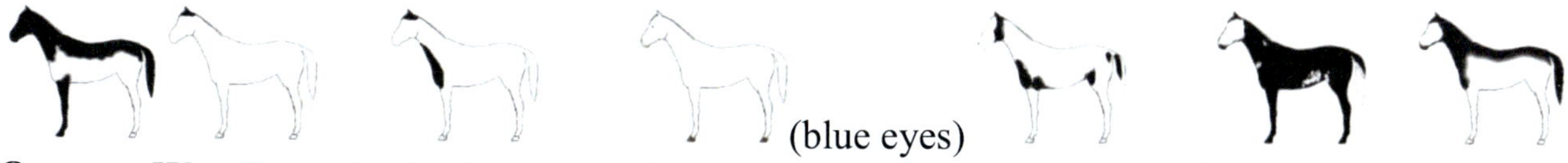

Overo   **War Bonnet**   **Medicine Hat**   **Extreme Tovero**   **Tovero**(Overo)   **Tovero**(Sabino))
                                            (blue eyes)   **Tovero**(Splash

Maximum Expression/Living Lethal

SABINO   SPLASH   TOBIANO   Morrocan Tobiano

Sabino   **Splash**   **Tobiano**   **Morrocan Tobiano**   **Blue Eye** or *Wall-Eye*

White (W)	No Color (ww)	Two Lethal Overo Genes	Rare Terminology	
White	Albino	Lethal White	Piebald	Skewbald

REORGANIZERS

Bends/Spots	Marble	Brindle	Lacing	Bird-catcher Spots	Camouflaging	White Striping	Tan Markings
Ben d'or *Smuts* *Grease Spots*	*Giraffe Markings*		*Lace* *Marbling* *Cob-web*		*Black Striping* *Zebra Stripes*		*(instead of white)*

FACIAL MARKINGS

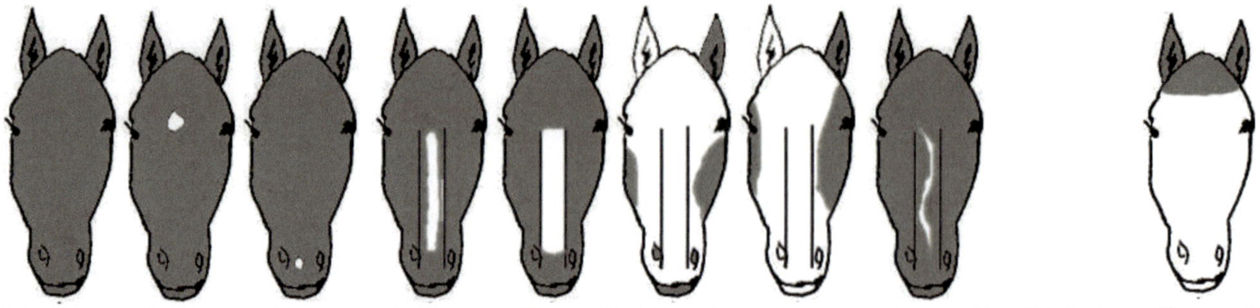

None	Star	Snip	Stripe	Blaze	White-Face Apron	Irregular Stripe	Bonnet
			Strip		*Bald-Face*	*Race (rare term)*	*War Bonnet*

LEG MARKINGS

| None | Coronet | Pastern | Fetlock | Sock | ½ Stocking | Stocking Leg | Ermine | Lightning Mark |

Whorls – (Clockwise = Intelligence, Counter-Clockwise = Not Smart)

 *Level Headed*    *Hi-Strung*   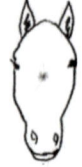 *Stubborn*

Made in the USA
Lexington, KY
13 December 2017